TODAY'S HEADLINES

12 THINGS TO KNOW ABOUT WILD WEATHER

BLACK RABBIT BOOKS | VICKI HAYES

Table of Contents

Satellites in space send information about weather back to Earth.

Think About It

Have you ever been in extreme weather? Maybe a big storm or a heat wave? How do you think we should cope with more cases of extreme weather?

Weather Can Become *Extreme*

1

Will it rain? Will it be warm? What should I wear? Weather is often the first thing people think about every day. Weather is the state of the **atmosphere** at a particular time and place. In most places, weather changes from season to season. It changes day to day. It may even change hour to hour.

The National Weather Service (NWS) is in the United States. The NWS gathers weather information. It uses **radar** and satellites. It also uses weather balloons. The balloons measure air temperature, pressure, wind, and **humidity**. All this information helps them make weather **forecasts**.

The NWS also issues weather watches and warnings. A watch means conditions might turn dangerous. People should keep a watchful eye on the sky.

A warning means dangerous weather is happening now. People should seek shelter and take other safety steps.

Sometimes the weather is extreme. Extreme weather could mean heat waves or high winds. It could mean blizzards or droughts. Extreme weather can damage buildings and roads. It can put people at risk. **Climate change** is causing more cases of extreme weather. Accurate forecasts can save lives and property.

4,000 Number of people working at the NWS.

Scientists who study weather are called meteorologists. • They use machines to predict weather faster and better. • In 2023, the United States had 28 weather disasters that cost over $1 billion each.

Weather balloons collect data that helps predict extreme weather.

A Thunderstorm Is an *Electrical Storm*

2

About 14.6 million thunderstorms occur worldwide each year. That's 40,000 thunderstorms a day. Thunderstorms bring much-needed rain to many places. They also produce electricity.

A thunderstorm forms when warm, moist air rises from the earth. As the air rises, it cools down. Thick clouds form. Cool air holds less water than warm air. If there is too much cool air, it forms **precipitation** and falls from the clouds. A thunderstorm has lightning, thunder, wind, and sometimes hail. Some **severe** thunderstorms can cause tornadoes.

Lightning is a giant electrical spark in the sky. It heats the air. The air gets five times hotter than the surface of the sun. Most lightning you see is a spark between the sky and the ground. Lightning can also travel

50,000 Degrees Fahrenheit (27,760 degrees Celsius) of an average lightning bolt.

Lightning strikes the Empire State Building in New York City about 25 times a year. • Lightning travels so fast, the moving air makes a sonic boom. That's what thunder is. • A lightning strike can cause a tree to explode.

Lightning comes before thunder because light travels faster than sound.

between clouds. Lightning often hits the tallest object in an area.

Lightning is very powerful. It can deliver enough electricity to light a 60-watt lightbulb for six months. It can cause **electrocution** in humans and animals. About 40 Americans die every year from lightning strikes. During a thunderstorm, people should stop outdoor activities like swimming. They should find shelter and wait until the storm passes.

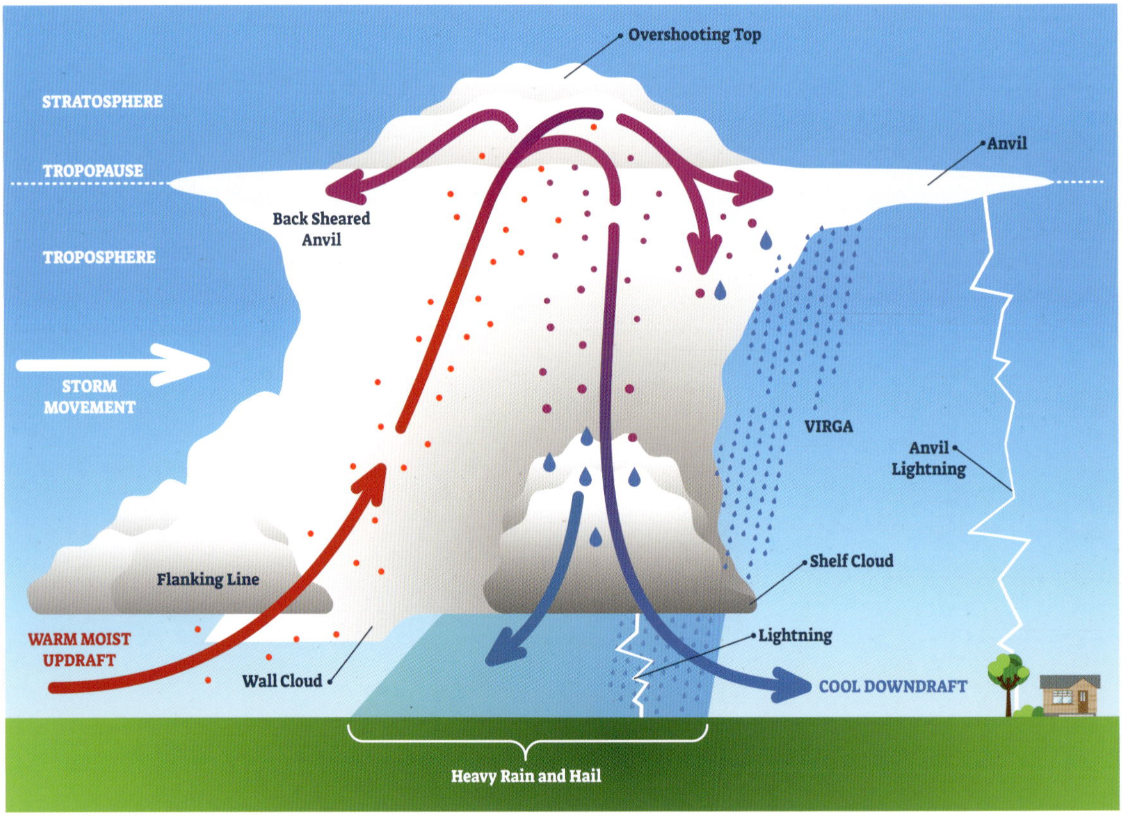

Tornado Winds Whirl at *Extreme Speeds*

3

Tornadoes touch down more often in the United States than in any other country. They begin as thunderstorms. The storm turns into a rotating column of air. This air extends from the storm to the ground. Tornadoes form most often during spring and early summer. They build up quickly. They blow through a small area. Then they die out after a few minutes to an hour.

Most tornadoes are weak. Small tornadoes have wind speeds between 65 and 85 miles (105–137 kilometers) per hour. Violent tornadoes can get up to 300 miles (483 km) per hour! Tornadoes can cause serious damage. They lift cars, animals, and people into the air. They can destroy everything in their path.

If a weather forecast predicts tornadoes, it is time to be alert. The sky may turn a dark greenish color. You may see a funnel cloud. You may hear a loud roaring noise. Seek shelter immediately. The safest place to be during a tornado is underground. Go down into a basement or other underground area. Some places have no basement. Then you should use a small, windowless room, such as a bathroom.

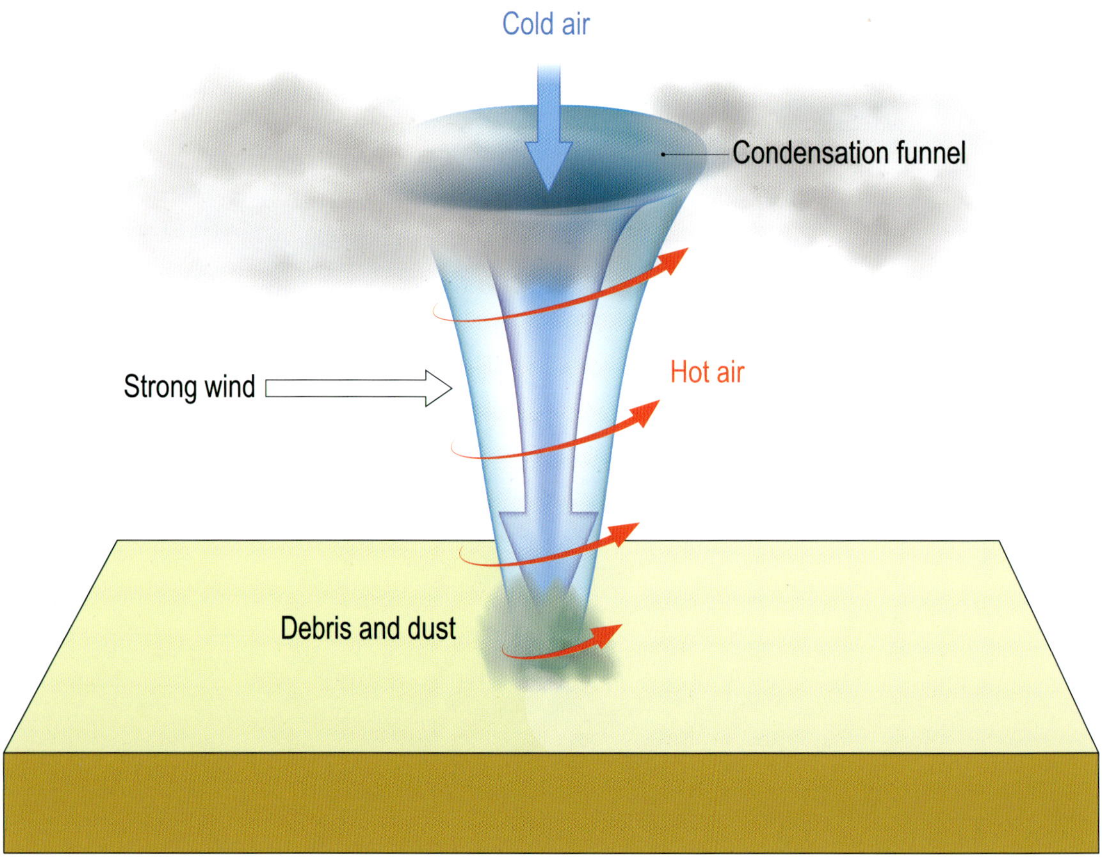

Tornadoes can form in all 50 states.

1,423 Number of tornadoes in the United States in 2023.

In 2023, tornadoes killed 83 people in the United States. • A tornado cluster is when several tornadoes occur in the same area over several days. Tornado clusters are increasing. • A waterspout is a tornado over water.

TORNADO ALLEY

Tornado Alley is in the central United States. It includes parts of states from South Dakota to Texas. Most US tornadoes have occurred in this region. But Tornado Alley is growing and moving due to climate change. It is stretching to Alabama, Tennessee, and Kentucky. People in these states are not used to having tornadoes. They will have to adjust to this wild weather.

4 Hurricanes Are Powerful *Tropical Storms*

Like tornadoes, hurricanes are powerful storms. But hurricanes cause 10 times as much damage. They are larger and last longer. Hurricanes can be hundreds of miles (km) across. Most last for a few days. Some last for two weeks. Hurricane season is from June to November.

Hurricanes are strong tropical storms. A tropical storm has winds up to 73 miles (117 km) per hour. When wind speeds get stronger, the storm is called a hurricane. A hurricane's winds can reach 200 miles (322 km) per hour.

Hurricanes form over warm ocean waters. Hurricanes tend to move westward. They get larger and stronger as they travel. But they start to die out once they reach cooler waters. Traveling over land also weakens a hurricane. When hurricanes hit land, they bring lots of wind and rain. One of the most dangerous effects of a

20,000 Number of US deaths from hurricanes between 1988 and 2019.

The Atlantic Ocean has 12 hurricanes a year on average. • Scientists believe climate change is making hurricanes stronger and more frequent. • Hurricanes in the Pacific Ocean are called typhoons.

Hurricanes produce heavy rains and strong winds.

hurricane is a quick rise in sea level. This is called a storm surge. The wind pushes ocean waters to shore. Storm surges cause flooding.

In the United States, deadly hurricanes strike coastal areas. The shores of the Gulf of Mexico are at risk. So is the Atlantic coast. Cities there are home to many people. And scientists have new concerns. Hurricanes are getting more frequent and more intense. They are also harder to predict.

Storm surges can develop quickly and cause heavy damage.

Tsunamis Are as Fast *as Jet Planes*

5

When underwater earthquakes happen, the ocean floor moves. This causes the water above it to move. The movement starts rolling waves. This series of large waves is called a tsunami. The word *tsunami* is Japanese for "harbor wave." Sometimes an underwater landslide or volcanic eruption can create a tsunami. The waves can be over 100 feet (30 meters) tall. They cause massive destruction when they reach the shore.

Tsunami waves start very small. They may only be 1 foot (30 centimeters) high. But the waves grow bigger closer to shore. They flow on shore as a sudden, violent flood. Often the first wave is not the largest. It may even take hours for the largest wave to arrive. Tsunamis can travel across the ocean at 500 miles (805 km) per hour. That is nearly as fast as a jet plane. Their

Like planes, tsunamis can cross oceans in hours.

impact can destroy buildings, trees, and most other structures. Many tsunamis develop with little warning.

In 2022, there was a volcanic eruption in the South Pacific Ocean. It created a large tsunami. The tsunami crossed the entire Pacific. It also affected the Atlantic Ocean and Mediterranean Sea. Damaged areas included Fiji, Hawaii, Chile, California, Japan, Russia, and New Zealand.

Volcán de Fuego, Guatemala

A drawing shows a tsunami hitting the coast of Hawaii.

167 Height in feet (51 m) of a tsunami that hit Sumatra, Indonesia, in 2004.

This is the deadliest tsunami in history. It killed 230,000 people. • When a tsunami reaches shore, it slows to 30 miles (48 km) per hour. • Tsunami waves from large earthquakes can continue to hit land for many days.

Fukushima, Japan, was hit with an earthquake and tsunamis in 2011.

JAPANESE EARTHQUAKE AND TSUNAMIS

On March 11, 2011, an earthquake struck off the coast of Japan. It was the largest in Japan and the third largest in the world. Residents got one minute of warning. Then the earthquake struck. Tsunamis followed 30 minutes later. The waves were 130 feet (40 m) high. They stretched 6 miles (10 km) onto shore. More than 18,000 people were killed.

Heat Waves Are *Getting Hotter*

6

Hurricanes and tornadoes are bad disasters. But more people die from heat waves each year than both of those combined. A heat wave is a period of high temperatures and humidity. Heat waves can last for days or even weeks.

Extreme heat can cause serious health problems. Heat exhaustion is one problem. It increases the heart rate and causes heavy sweating. A person can become dehydrated. Left untreated, heatstroke occurs. This is when the body's temperature gets too high. It is life threatening.

Cities are especially at risk for heat waves. Asphalt and concrete surfaces store heat. At night, this heat is released. It causes nighttime temperatures to increase. This creates an urban heat island.

Heat waves are becoming more common. They are also lasting longer. Meteorologists agree that global warming is the cause. Earth's temperature is rising. In 2022, 1,714 people died in the United States from heat-related causes. The summer of 2023 was the hottest on record. Deaths in many areas were 50 percent higher than in 2022. Even the Arctic had its warmest August ever.

Rising temperatures affect the glacier habitats of polar bears.

Record high temperatures are recorded every year.

70,000 Number of people in Europe who died during a heat wave in 2003.

Heat-related deaths in the United States are up 95 percent over the past 14 years. • In the heat wave of 2023, 334 people died in Texas. • In 2023, Arizona had 31 days in a row with temperatures above 110 °F (43 °C).

Droughts Are *Slow Killers*

7

A drought is a long period with no rain. Droughts can happen anywhere in the world. But they are more common in dry areas. Droughts are slow to begin and end. They can last a few weeks or a few years.

Droughts do not occur suddenly. Many are caused by changes in air patterns. The air might move water vapor in unexpected directions. Then areas expecting rain will get drought. Droughts have other causes too. Deforestation and poor farming practices can lead to drought. So can changes in ocean temperature. Climate change is another big cause.

Plants and vegetation dry up with little to no rain. Soil becomes drier. Wildfires increase. There can even be flash flooding. The land is too dry to absorb water. Surface water sources, like ponds and streams, dry up.

Droughts can make it difficult for animals to find food and water.

160 million Number of children living in areas with high levels of drought in 2024.

Droughts have caused 70 percent of all deaths from natural disasters. • The top driest years for the United States were 2012, 2022, and 2023. • Droughts in 2023 cost the United States $14.5 billion.

Some places may lack clean drinking water. This leads to a spread of diseases. A lack of water harms humans and animals. It also damages crops. This can lead to **famine**.

Drought and famine have been severe in Africa. Ethiopia had a drought for three years. Then in 2023, it rained. There was flash flooding. Homes and buildings were washed away. Around 300,000 people were affected.

A family crosses a river during flash flooding in Ethiopia.

Dust Storms Can Move *Across Oceans*

8

Dust storms look like giant, dark, cloudy walls. They are blown by winds from thunderstorms. The walls of dust can be 1 mile (1.6 km) high or more. They can be hundreds of miles (kilometers) long.

Dust storms happen in various places. Many start in deserts in Asia and Africa. These storms can hold large amounts of dust. Winds can blow the dust across the Atlantic Ocean. The dust can reach the Caribbean and eastern United States.

But dust storms don't always start in deserts. They can start in any dry, hot area. Loose dirt or sand is picked up by strong winds. The US Southwest often has dust storms. All the dust and debris can be very dangerous. It is hard to see far distances. It makes driving and flying dangerous. The dust causes poor air quality. It triggers breathing problems and other health issues.

Dust storms are most common in spring and summer.

60 Average wind speed in miles (96 km) per hour of a dust storm.

A dust storm is also called a haboob. • In May 2023, an Illinois dust storm caused a 72-car pileup. • A dust devil is a small, short-lived whirlwind. It may cause minor damage.

THE DUST BOWL During the 1930s, the US Great Plains suffered a long period of drought. There were many dust storms. The whole area became known as the Dust Bowl.

Dust storms are becoming more common. Water is limited. Animals **overgraze**. Forests are cut down. And climate change brings more droughts and hot temperatures. Desert dust has doubled in the 20th century. All this makes conditions ripe for more dust storms.

A massive dust storm hit the area on April 14, 1935. This storm was known as the Black Blizzard. About 100 million square acres (40 million square hectares) were affected by the Dust Bowl.

A huge dust storm blows across the Arizona landscape.

Wildfires Burn Large Areas *Quickly*

9

Wildfires are an important part of nature. They remove dead or decaying matter. They kill off diseased plants and harmful insects. They burn thick canopies of leaves. Then more sunlight can get through to new plants.

Climate change has caused wildfire patterns to change. The fires are more frequent. They are more intense and larger. Wildfires are extremely destructive. They cost billions of dollars to control. They burn buildings and crops. They harm people and animals. In 2020, 17 million animals were killed by wildfires in Brazil.

Wildfires, also called forest fires, start in forests or brush areas. Some wildfires are caused by lightning strikes. Most are caused by humans. Wildfires can happen anywhere. They are common in the forests of Canada and the United States. In southern California, they are more intense. This is due to hot, dry Santa Ana winds. Wildfires can last for weeks. They can burn across thousands of acres of land.

FEEDING THE FLAMES

Three elements must be present for a wildfire to start: fuel, oxygen, and a heat source. Dry weather and drought turn plants into fuel. Wildfires spread quickly over land. Strong, dry winds feed the fire oxygen and spread the embers farther away.

Helicopters try to control wildfires by spreading water.

70,000 Number of wildfires each year in the United States.

Humans cause about 85 percent of all wildfires. • In 2022, 7.5 million acres (3 million ha) of US wildland were destroyed by wildfires. • In 2019, wildfires in Australia burned for six months straight. They destroyed 46 million acres (18 million ha).

Aftermath of 2022 wildfires in Dadia National Park, Greece

Blizzards White Out *the Landscape*

10

Blizzards are severe winter storms. Temperatures drop and snow falls fast. Winds over 35 miles (56 km) per hour blow the snow. This creates a whiteout. It is hard to see in these conditions. People could get lost and become stranded. People caught in the storm could get **frostbite** or even **hypothermia**. In many cases, people have died from the cold.

A blizzard forms when a mass of cold air moves out of the Arctic or Antarctic. In the United States, blizzards usually occur between December and March. Blizzards can be extremely dangerous. Roads become slick and covered in snow. Driving is difficult and accidents are common. Airports and railroads usually close. Winds create big snow drifts that block roads and doors. Blowing snow can drift over cars and even buildings.

50 Depth in feet (15 m) of snow drifts in the East Coast after the March 1888 blizzard.

More than 400 people died in the 1888 blizzard. • There were 12 blizzards in the United States in the winter of 2021–22. That is a record! • The snowstorms in Texas in 2021 caused $295 billion in damage.

Blizzards often create dangerous road conditions.

Climate change affects winter storms. Warm air traps water vapor. When temperatures do drop, snowfall is heavy. Changes in air patterns are sending winter storms farther south. In 2021, there were record-breaking snowstorms in Texas. They lasted for almost nine days. There was snow, sleet, and freezing rain. Roads became icy. One highway had a pileup of over 100 vehicles. The temperatures were extreme. The wind chill was below zero. Water pipes burst. Stores ran out of food. Millions lost power.

Think About It

Light, fluffy snowflakes look harmless and even beautiful. In what ways can snow be dangerous?

Ice Storms Take Out *Power Lines*

11

An ice storm is a heavy downpour of freezing rain. The rain freezes when it hits the ground. Roads, sidewalks, and trees become covered in ice. This ice coating is called glaze. Most of the glaze is less than an inch (2.5 cm) thick. But this small amount is dangerous.

Glaze-covered streets and sidewalks make travel difficult. The glaze makes these surfaces very slick. People walking may slip and fall. Bridges and overpasses freeze first. Vehicles can lose control and spin out. Clear ice is hard to see on the black surfaces. This is called black ice. It is a big problem on highways.

Glazed tree branches and power lines are especially dangerous. They can get up to 30 times heavier than normal when covered with glaze.

Signs warn drivers where ice may form quicker on the road's surface.

This can cause them to collapse, especially with high winds. Then people lose power.

Ice storms are very damaging in the southern United States. This area usually has a warmer climate. Buildings, crops, and people are not prepared for cold weather. In 2021, snow and ice knocked out power for 4.5 million homes in Texas. More than 100 people died.

240,000 Number of buildings that lost power during an ice storm in January 2024.

The storm covered Oregon cities in a thick layer of ice for days. • A half inch (1.3 cm) of ice on a power line can add 281 pounds (127 kg). • There was a devastating ice storm in January 1998. It affected millions of people in the northeast United States.

Think About It Losing power at home is bad. But think how these places would be affected by a power outage: hospitals, roads, schools, and airports.

The 1998 ice storm froze over areas of southeastern Canada.

Communities Work to Prepare for *Wild Weather*

12

Every year, extreme weather kills around 90,000 people worldwide. The cost of damages is also high. Countries are taking steps to prepare for extreme weather. The United Nations has started a program called Early Warnings for All. It aims to create warning systems for all countries by 2027. Early warnings can reduce losses and damages. It can save lives.

Wild weather costs billions of dollars in damage each year.

Building better **infrastructure** can also help. Coastlines clear space to protect from flooding and sea rise. Cities plant more green spaces to cool heat waves. Farms develop crops that can withstand high heat, drought, and flooding. Engineers update or build structures that can survive extreme weather.

In Florida, a new town is prepared for hurricanes. Babcock Ranch opened in 2018. It was built 20 miles (32 km) inland. It has a strong drainage system.

The buildings are 30 feet (9 m) above sea level. The power lines are underground. The town is America's first solar-powered town. The solar farms have large batteries that keep power on during blackouts. Babcock Ranch is called the "hometown of tomorrow." Building smart infrastructure like this can keep people safe from future wild weather.

Workers install solar panels on the roof of someone's home.

$830 billion Money from federal grants that will help protect US transportation systems from climate change.

Grants were approved in 2024 for 80 projects nationwide. • Projects include roads, bridges, highways, ports, and public transportation. • The projects will improve draining, move roadways, and lift bridges.

AMERICAN EXTREMES The United States has more weather disasters than any other country. This is due to its geography. The United States has oceans, mountains, and a gulf. It also has a clash of air masses. Cold air blows down from the Arctic. Warm air comes up from the tropics. They often meet over the United States.

Babcock Ranch, Florida, utilizes solar panels on building rooftops.

How to Prepare

Be Informed

Know what kind of extreme weather may occur where you live. Check the forecast before leaving the house. Is a storm heading your way? Watch your local news stations too. They will let you know if there are any watches or warnings in effect.

Make a Plan

Talk with your family. Choose a safe place in your house to stay during big storms. Decide what you'll do if there is a tornado, wildfire, or hurricane. What will you do if the power goes out? Planning ahead can make events less stressful. It can save time and keep you safe. Make sure to practice your plan too.

for Wild Weather

Prepare an Emergency Kit

Fill it with food, water, and flashlights. You also want extra flashlight batteries and medical supplies. Include copies of important documents too. Visit www.ready.gov/kit for a list of more supplies. Keep your kit someplace easy to grab. Then you'll be prepared to grab it quickly in an emergency.

Prepare Your Home

Trim branches that are close to your home. They could break off in a storm. Close windows and put away outdoor items before a storm starts. Make sure outside doors are secured shut.

Glossary

atmosphere
The gases that surround a planet.

climate change
A long-term change in a region's typical weather patterns.

electrocution
The killing of someone by electric shock.

famine
A very great shortage of food that affects many people over a wide area.

forecast
A scientific prediction of what the weather will look like in the future.

frostbite
A condition in which part of someone's body freezes or almost freezes.

humidity
The amount of moisture in the air.

hypothermia
A physical condition in which the temperature of the body drops dangerously low.

infrastructure
The basic equipment and structures that are needed for a country or region to function properly.

overgraze
Allowing animals to graze until they damage the plants.

precipitation
Water that falls to the earth, such as rain or snow.

radar
A device used to locate precipitation, calculate its motion, and estimate its type (rain, snow, or ice).

severe
Intense, extreme, or dangerous, when describing weather.

For More Information

Books

Borgenicht, David. *The Worst-Case Scenario Survival Handbook for Kids.* Philadelphia: Quirk Books, 2024.

Carson, Mary Kay. *The Tornado Scientist: Seeing Inside Severe Storms.* New York: Clarion Books, 2023.

Kerry, Isaac. *Climate Change and Extreme Weather.* Minneapolis: Lerner Publications, 2023.

Websites

Kids Discover
https://online.kidsdiscover.com/unit/extreme-weather

Tree House Weather Kids
https://web.extension.illinois.edu/treehouse/rockweather.cfm?Slide=1

Weather Wiz Kids
www.weatherwizkids.com

About the Author

Vicki Hayes has a master's degree in early childhood education and more than 30 years of experience teaching children of all ages. She currently works as a teacher and writer.

Index

• Top Rank is an imprint of Black Rabbit Books. • Edited by Alissa Thielges | Designed by Danny Nanos • Photographs © Alamy Stock Photo/North Wind Picture Archives, 18-19; Dreamstime/Ahaze, 38, Chrisgandy, 37, Dezzor, 4-5, Dmitrii Afanasev, 16, Dirk Ercken, 37, Gary Blakeley, 36, John Sirlin, 28-29, Lane Erickson, 35, Mykhailo Polenok, 7, Realitytimes, 11, Smallcreativeunit5, 20, Starzon450, 22, Sue Harper, 45, Waingro, 2; Getty Images/Johner Images, 24, NurPhoto, 32-33; Shutterstock/Achira22, 31, Ahanov Michael, 45, Alex Photo Stock, 44, Belinda Pretorius, 48, BEST-BACKGROUNDS, 14, Bilanol, 42-43, Designua, 12, Dushlik, 17, Everett Collection, 27, FotoKina, cover, gobigoo, 6, Grey Zone, 13, Janice Storch, 21, Jolygon, 46-47, Kedardome, 34, leolintang, 23, Liz.dee, 18, meunierd, 39, moreimages, 41, Nikolay Zaborskikh, 2-3, 8-9, photoschmidt, 15, Pixel-Shot, 44, Piyaset, 25, Rosemarie Mosteller, 27, Stanley Dullea, 26, Toa55, 31, Todor Stoyanov, 40, TSViPhoto, 40, VectorMine, 10 • Printed in China

Library of Congress Cataloging-in-Publication Data Names: Hayes, Vicki C., author. | Title: 12 things to know about wild weather / by Vicki Hayes. | Description: Mankato, MN: Black Rabbit Books, [2025] | Series: Today's headlines | Includes bibliographical references and index. | Ages 9–13 | Grades 4–6 | Identifiers: LCCN 2024020337 | ISBN 9781645823919 (library binding) | ISBN 9781645824138 (paperback) | ISBN 9781645824350 (ebook) | Subjects: LCSH: Climatic extremes—Juvenile literature. | Classification: LCC QC981.8.C53 H39 2025 | DDC 551.55—dc23/eng/20240510 | LC record available at https://lccn.loc.gov/2024020337